Edward H. Dixon

The Kidney

Its Structure, Functions, and Diseases, Bright's Disease

Edward H. Dixon

The Kidney
Its Structure, Functions, and Diseases, Bright's Disease

ISBN/EAN: 9783337173371

Printed in Europe, USA, Canada, Australia, Japan.

Cover: Foto ©berggeist007 / pixelio.de

More available books at **www.hansebooks.com**

THE KIDNEY

ITS

STRUCTURE, FUNCTIONS, AND DISEASES

BRIGHT'S DISEASE

THE URINE; ITS CONSTITUENTS; CHEMICAL TESTS FOR THE
VARIOUS DISEASES; THEIR SYMPTOMS
AND TREATMENT

ADAPTED TO THE POPULAR COMPREHENSION

BY

EDWARD H. DIXON, M. D.

NEW YORK

J. S. REDFIELD, PUBLISHER

140 FULTON STREET

1871

EDWARD O. JENKINS,
PRINTER AND STEREOTYPER,
No. 20 North William St.

INTRODUCTORY.

The reader will soon discover that this little book is designed only for his instruction; not to advertise any specific to "cure" any of the diseases of which it treats. The uniformity of the laws of the human organism in producing disease when disobeyed, and the certainty with which the same laws aid our efforts for its removal when the error is discovered in time, and the body placed in such a condition as to permit their unimpeded action, is the first requirement in knowledge of the scientific physician. This great truth is admirably illustrated by the structure and functions of the kidney. Its organization is extremely delicate. It is the great *Purifier* of the body. If its function be entirely suspended for a few days, the person dies, blood and brain poisoned by the urea. Medicine can have no effect whatever.

Acute inflammation of the kidney may kill in a few days, or end in Albuminuria or Bright's Disease. It is caused by a sudden chill or wet feet. Yet a few blankets and an alcoholic vapor bath in time, will often cure it. They are far preferable to any medicine whatever. Here is the disease and the remedy—caused and cured by disobedience of, and obedience to, an organic law. In regard to the more obscure diseases, a critical investigation

(3)

into the habits, diet, and place of residence of the patient, and a careful chemical and microscopic analysis of the urine, will often aid us in giving the patient advice and chemical remedies that will rectify the abnormal conditions of the secretion, and with the beneficent aid of the organic law, restore him to health. But the reader should never forget that medicine alone will not "cure" disease; it can only aid nature; and whoever desires to give intelligent aid must diligently study and reverently obey her unerring laws.

It was the author's intention to publish this pamphlet in the same form as the Scalpel, and his other essays and lectures. But our excellent friend, Mr. Redfield, advised this, as less scholastic and repulsive to the general reader. We hope it will repay his efforts for its diffusion, and save many readers from the designs of the quack.

EDWARD H. DIXON,
42 *Fifth Avenue.*

THE KIDNEY,

ETC.

BRIGHT'S AND OTHER DISEASES OF THE KIDNEYS.

In the year 1827, Dr. Bright of London, first published in his select report of Medical Cases, his investigations into several morbid conditions of the kidneys, accompanied with dropsy and the appearance of albumen in the urine. Previous to that time, dropsy when it became the last and fatal symptom of disease existing in other internal organs as well as the kidney, was ordinarily viewed as the disease itself that had caused the death. It was a very common enquiry amongst patients and no small number of physicians, whether the patient had abdominal dropsy, or dropsy diffused throughout the general tissue of the body, and what remedies were adapted to " carry it off." It was certainly well known, at that time, and long before to pathologists, that enlargement of the liver, the spleen, and of the kidneys also from chronic inflammation, would produce dropsy of both varieties ; but no one had ever observed a generally fatal disease of the kidneys, the distinctive symptom of which was such an amount of albumen in the urine, as in a few months to waste away the

body, and destroy life by starvation. This it
always does, as albumen is the chief constitu-
ent of the blood, and the substance of which
the body is chiefly formed ; from this peculiar
symptom, the disease was called, Albuminuria, which is its proper scientific name,
although, in compliment to its discoverer it
was originally called " Bright's Disease." We
very well remember, when a student of the
late Dr. Valentine Mott, in 1830, quite a rivalry
existed amongst the surgeons of the city, who
should demonstrate to the class of the Rut-
ger or Barclay Street Colleges, the first
specimen of the disease ; it was ended by the
exposure to the class attending the New York
Hospital, of two specimens of the diseased
kidneys taken from the body of a patient by
the late Dr. John Kearney Rodgers, who was
then one of the surgeons of that institution,
and who courteously brought them for our
inspection. It was evident to all who examined them, that when characterised in as mark-
ed a degree, as these specimens were, the disease would admit of no remedy ; the beautiful
structure of the kidneys was quite disorgan-
ised ; and, but little of it distinguishable. At
that time nothing was known of the different
varieties of the disease, and still less of the
causes. Since then it has been the subject of
critical research by the English, French and
German Pathologists, and it is now very well
understood.
    But it is necessary to give the reader a
comprehensive, though necessarily a short
sketch of the organic laws which govern the
kidneys and their secretions as well as the

structure by which they separate from the blood, urea and other salts hurtful to the body, and then he will be able better to understand the disease of the organ which causes albuminuria, and the means to be used to check its progress to its most frequent consequence—death by blood starvation. It will be observed in this, as well as all our other articles, we address the intelligent reader only. It is by logical use of the knowledge derived from the laws which govern the structure of the kidneys in health that he can avoid the disease, or stop its progress; not by swallowing the compounds of the unprincipled quack, as buchu, lithia, and other nostrums constantly advertised in the newspapers. We would have the reader remark, that these remedies are only very slight aids in the hands of the most skillful physicians, often producing no benefit whatever when they contain the genuine article, by the name of which the vender of the nostrum imposes upon the credulity of the afflicted.

There are two processes continually at work in the human system, upon the healthful action of which depend the health and existence of the being. The first is *nutrition*, or the supplying the body, from the external world, with those materials which enter into its composition, and with which it is necessary to be provided to maintain its life. The other process is that of *elimination*, or the casting off from the body those particles for which it has no further use, and which, if allowed to remain, would result in disease and the total disorganization of the system.

The *blood* is the great nutritive principle. Blood consists of all the chemical elements contained in the tissues. This is conveyed to them by the heart and arteries, and from it is deposited in them all the elements required for their formation. In exchange, is returned such parts as, having performed their duty, are cast off by nature as useless. Thus we have the process of *assimilation*, and this plastic force is perfect when in health, as is shown by the absorption of the virus in vaccination, and by the formation of eschars after the healing of wounds. Parts having once changed their condition remain so, though still having a tendency to return to the normal state. We have a fine illustration of this law in secondary syphilis and its promulgation. When a tissue remains in the normal state as regards its capacity for action, the process of nutrition and elimination go on in the regular manner, and the part remains of the same size and power; but bring the tissue into a little more action, and we see a change; there is an increase of nutrition, and consequent bulk—*hypertrophy*. This results from more particles being added to the tissue to compensate for the larger amount of power required. This is beautifully illustrated in the increased size of the arm of the blacksmith and the legs of the dancer.

We have the reverse of this in *atrophy*, or diminution of size. This results from a decrease of nutrition, caused by want of action, or obstruction to the passage of blood through some important blood-vessel. Thus, when an arm is paralyzed, its action is less: it becomes *atrophied.* As a general rule we can reverse the laws of hypertrophy, and apply them to atrophy. We find examples where this process resembles diseased action ; thus from some obstruction to the passage of food from the stomach, according to the above law, the pyloric, or upper orifice of that organ may become hypertrophied, and may resemble cancer. We also find examples of this action in the kidney, its fellow having been destroyed ; and in the heart and bladder in diseases peculiar to them ; yet the independent process of hypertrophy or increase of bulk, is strictly a healthy one.

A law has been established to this effect, that increase of function leads to augmentation of bulk, but exception is given in case of the organs of special sense. Thus one eye does not become hypertrophied when the other is blind, etc. But it seems that the law is at least true of all muscular tissue, though not of the nervous. There is an exception to this rule in the nerves of the uterus, when increased during pregnancy, as has been demonstrated ; also that the increase of nervous substance causes increase of susceptibility. In the organs of special sense, the nerves are the principal agents, (the voice being an exception). The muscle of one eye would not be brought any more into play on account of the other being destroyed. But the nerve will get a greater degree of susceptibility to impression by increased activity ; thus the eye of the mariner can discern objects at a greater distance than a person less accustomed to this mode of observation.

As the nutritive process is providing the tissues with new material, there is another force in operation, that of *elimination*, by which particles are removed as new ones are added. The disintegration or decay of particles is founded upon a law " that every particle of the body is formed for a certain period of existence in the ordinary conditions of active life ; at the end of which period, if not previously destroyed by outward force or over exercise, it degenerates, and is absorbed or dies, and is cast off." This constitutes the process of *secretion* and *excretion*. The fœces and urine are excretions.

These particles, if not cast directly off, as by the skin, are absorbed, and again show themselves at the various outlets of the body : namely, the lungs, skin, bladder, and rectum. This law has an exception, for we find that those parts which have taken an active position and become foreign matter by death may be atrophied, but do not disappear, provided they are not interfering with other parts or functions. But if this interference takes place, they are absorbed or cast off—absorbed as are the fangs of the milk-teeth, the crown being cast off, as the hair from its little gland, to make way for a new one. By these actions it is stated that the whole substance of the animal economy is renewed during seven years.

One word in regard to absorption. Some membranes pour out their secretion, and through defect of the absorbing apparatus, the effused fluid is not removed, thus constituting dropsy. If now the blood-vessels are emptied of some of their contents, either directly, as in bleeding, or by the use of cathartics, diuretics, etc., their loss is replaced by absorption and the simple dropsy is stopped for a time ; sometimes if the general hygienic conditions are favorable, cured.

Absorption is in inverse ratio to the fullness of the blood-vessels. According to this law, we see the cause of effusion in tardy or restrained circulation.

As man is provided with a perfect organization, we find that when the proper functions of one organ are interfered with by disease or otherwise, nature strives through other sources to compensate for the deficiency resulting from its non-action. Thus there is a perfect balance of action in the system, and nowhere is this better seen than in the secretions.

Thus, when one of the secretions is diminished, the others are increased. For example, in warm weather or from violent exercise, the exhalation from the skin is increased, and in consequence, the quantity of urine passed is decreased. Again, during cold weather, as the functions of the skin are not so active, the quantity of urine is increased.

In general the amount exhaled from the skin, and that passed in form of urine is nearly the same. Urine arises from various *sources*. Food is introduced into the stomach, and by the process of digestion and absorption passed into the blood for the nutrition of the tissues: and, as we have seen, a dead particle is removed, and at the same time replaced by a new one. The blood may contain some elements which it has received from the food, and which it finds the tissues have no use for; these elements, together with the dead particles, are removed from the system by the secretions. Any element not wanted for the use of the body, and which exists in a state of solution, always passes away from the system by way of the kidneys.

If large quantities of liquid are taken into the stomach and absorbed, the kidneys act as pumps to remove the excess of fluid from the system.

The secretions constituting the sewerage of the body, and a certain quantity of refuse matter being obliged to be removed to insure life, we can readily see the amount of general derangement to the whole system, resulting from the blocking up of one of these channels of exit. Also that we may imagine a disease of the secreting organ from the fact of the secretion being changed from the normal standard, when in fact the secreting apparatus may be performing its healthy functions and removing from the system such matter as results from the disease of some remote organ.

We should therefore be careful in a proper examination of the secretion and the accompanying symptoms, to enquire into the food and habits of the person, which tend to alter materially the secretions, and especially the urine.

Intemperance in the use of spirituous liquors is one of the great sources of urinary disease, so is in fact, any kind of debauchery which tends to reduce the vital power of the system. Bad air, unwholesome food and locations, tend to produce disease of the kidney. Onanism, venereal excesses and the abominable habit of partial sexual intercourse to avoid offspring are fruitful causes; but we must not anticipate the causes and treatment before we describe the

Anatomy of the Kidney. The organs or glands which secrete the urine are the kidneys, two in number, situated in the region of the loins, on each side of the spinal column, in form resembling a kidney bean. Instances have been known where there existed a third kidney, situated generally in front of the spinal column. Occasionally, also, there is but one gland. The total absence of these organs has been observed in the fœtus. Each kidney is about four or five inches in length, two and one half in breadth, a little over one inch in thickness, and about five ounces in weight. The dimensions and weight of course vary in different individuals. The kidneys are of a reddish brown color and of an extremely fragile texture, so much so, that they are liable to be ruptured by any violence applied to them directly or indirectly. They are generally surrounded by a quantity of fat.

The kidney is enveloped by a fibrous covering, which can be readily detached from the surface of the gland. If now we should make a longitudinal section of the organ, we should find it to be composed of two substances, differing in several particulars. The external or vascular portion is a structure of a red dusky color, composed of blood-vessels and convolutions of minute tubes $\frac{1}{480}$ of an inch in diameter. Throughout this portion of the kidney are scattered a multitude of globular bodies of a red color and about the size of a pin-head $\frac{1}{100}$ of an inch in diameter. This is the portion which secretes the urine. The minute globular bodies above described, secrete the water which in passing through the fine tubes of this portion obtains its solid constituents.

Still deeper in the gland and in immediate connection with the external portion, is the *tubular portion*, composed of fine tubes about the size of a hair, and arranged in the form of cones, of a rose color. The object of these tubes is to convey the urine secreted by the external portion to a large interior cavity shaped like a funnel. From this interior cavity of the kidney, and continuous with it, is a tube about the size of a goose-quill called the *ureter*, which serves as a duct to convey the urine to the bladder.

The *urinary bladder* is an oblong sac, constricted at one end, forming its *neck*. This neck is encircled with the prostate gland, which in old age and sometimes in youth, from sexual excesses and self-abuse, is found hypertrophied. The walls of the bladder are composed principally of muscular tissue, which serves by its contraction, to expel the urine. Continuous with the bladder is the *urethra*, a canal which conducts the urine from the body, which, in the male, passes from the neck of the bladder to the extreme termination of the penis, and is about nine (9) inches in length.

We can thus trace the urine from its secretion in the external portion of the kidney, its passage through the fine tubes of the internal portion to the large interior cavity of that

gland, its course through the ureter to the
bladder, where having accumulated in sufficient quantity to cause distension, it is finally
expelled from the body by way of the urethra, which expulsion is caused in part by the
contraction of the muscular portion of the
walls of the bladder.

The urine is an amber-colored fluid. The
density of color will depend upon the health,
diet, and period at which it is passed. The
intensity of color is in inverse ratio to the
quantity passed; when the urine is scanty it
is high colored, when in large quantity it is
the reverse. Thus in diabetes where the
quantity passed is immense, the urine is often
free from color. Again, some substances introduced into the stomach as food or medicine,
produce an effect upon the color of the urine
—for an example, rhubarb. The change of
color can not in all cases be relied on as an
evidence of disease.

The odor of urine is peculiar. As in regard
to color, so is it with odor; it is altered by
some articles of diet, as asparagus. The odor
is liable to change from disease. In diabetes
it has the smell of hay, in nervous diseases
aromatic, in injuries of the spine that of ammonia; it soon becomes putrid when, from
some disease of the urinary organs, the urine
contains pus, mucus, etc.

It is, when voided, of the same *temperature*
of the body. Its *taste* is salt and bitter during
health, but in disease it is changed; thus in
diabetes, from the large amount of sugar in
solution, the urine is sweet. Healthy urine is
acid. That passed after eating is less acid and

may occasionally be found alkaline, while that passed at the period most distant from the meal is most acid.

The quantity of urine secreted in twenty-four hours is generally from thirty to forty ounces. This, of course, depends upon the diet and habits of the person and amount of liquid taken. Thus a man whose normal quantity was in twenty-four hours thirty ounces, passed fifty-six ounces after swallowing a quart of water in the course of the day. The habits of the French are such from the use of wine, that they pass about forty-three ounces for the men and forty-seven ounces for the women. The quantity of urine is changed from mental excitement; thus a young woman who naturally passed in twenty-four hours thirty-five ounces, voided eighty-six ounces after an hysterical fit. It is also changed in disease; as in the case of diabetes it is increased to many quarts, and decreased in inflammatory diseases of the kidney.

The limpid urine of health may, by the admixture of pus, mucus, etc., be rendered so *consistent*, that it may be drawn into threads. The *density* of urine depends upon the quantity secreted. The smaller the quantity the greater the density. It is different in the two sexes, the density being the greatest in males and at the prime of life; its density decreasing at childhood and old age. The specific gravity of urine in health varies from 1015 to 1025.

Urine upon being allowed to stand, soon decomposes. Clouds form in it and sink to the bottom, afterwards decomposition is an-

nounced by an unpleasant odor. As decomposition advances, the odor becomes more
disagreeable, and a mold forms on its surface.

Urine consists of water with various organic
and inorganic substances held in solution.
We have seen that the quantity of urine
secreted during twenty-four hours is on an
average thirty to forty ounces; this will be
found to contain from six hundred to seven
hundred grains of solid matter, although this
amount is not constant, being changed by
character of food, disease, etc.

CHEMICAL COMPOSITION OF URINE.

Water, ...	950
Urea, ...	25
Uric acid, ...	1
Fixed salts, . . .	14
Organic matter, . . .	10
Parts, . .	1000

The fixed salts are chlorine, sulphuric acid,
phosphoric acid, potash, soda, lime, and magnesia.

The most important of these constituents
are water, urea, and uric acid.

There is also a trace of sugar in healthy
urine.

Urine in disease. As the object of the
urinary secretion is to free the body from all
refuse matter, either arising from nutrition or
the disorganization of tissue — otherwise the
wear and tear of the body — any derangement
of these functions will produce a correspond-

ing alteration in the urine. Some of the normal constituents of the urine may be in excess; there may be a deficiency of one or more of them, or some of them may be absent altogether. The urine may contain substances not found in it during health.

If, after the urine is voided, a precipitate fall to the bottom, it is called a *sediment;* if the substance is precipitated in the bladder or kidney, it is called *gravel;* and if it accumulates and concretes in large masses in any of the urinary passages, it constitutes *stone.*

The urine is an important indicator in our discrimination of the various diseases of the urinary organs, and too much attention can not be given to an examination of that secretion, in case of derangement of those organs.

We propose to give the chemical tests necessary to detect the various constituents in health and disease, with a description of their properties, their relations to a diseased state of the system, and the indications of their treatment.

The specimen of urine to be examined should be that voided in the morning on rising.

Urea is a colorless substance, of a crystalline nature, existing in the healthy urine in proportion of about 14 parts in 1000, being often in disease more than double that quantity. Urine containing an excess of this element, has a high specific gravity from 1030 to 1050.

The specific gravity of urine is obtained by means of an urinometer. This is a small instrument consisting of a glass stem graduated;

at the one extremity is a bulb, containing mercury. To obtain the specific gravity of a given specimen of urine, it is only necessary to float this little instrument on its surface, and the point on the graduated stem, corresponding to the surface of the urine, will give the specific gravity.

Test for Urea. Put a little urine in a watch-glass, and evaporate over a spirit lamp about one-half its bulk; then add an equal quantity of pure nitric acid, and evaporate to dryness in a cool place. Delicate crystals of nitrate of urea will be formed. Nitrate of urea is soluble in about eight times its weight of cold water, but is more readily dissolved in hot water. It is also soluble in alcohol.

The microscope is a valuable adjunct in the examination of urinary deposits. It, however, requires care, experience, and an extended series of illustrations.

Uric acid, in the pure state, is a colorless, crystalline substance, forming about 0.4 parts in 1000 in healthy urine, and in disease as high as two parts. When in excess, the urine has a reddish brown color, and upon being left to stand throws down a red or yellow sediment. Its specific gravity is rarely above 1020. It has an acid reaction.

Urine is tested, as regards its acid or alkaline properties, by means of blue litmus paper. If a piece of this paper be immersed in urine containing an excess of acid, it will turn red. If this reddened litmus paper be wet with an alkaline solution, the blue color will be restored.

A deposit of uric acid may occur from the

fact that the urine is so concentrated, that there is not enough water to hold the acid in solution. The sediment is insoluble in alcohol, and nearly so in water. It can be dissolved in a strong solution of caustic potash.* *On heating the urine, the sediment is not re-dissolved.*

If uric acid be not deposited as a sediment, as above stated, if it exist in the urine, it may be shown by adding to four ounces of the urine, one and a half drachms of hydro-chloric acid. If the vessel be covered and left for a day or two, a red or yellow sediment will be deposited.

Urate of Ammonia. Urine containing an excess of urate of ammonia is generally high-colored and turbid when voided. Upon being allowed to stand, it deposits a red or pink sediment, which is not crystalline in its nature. *By heating the urine, the sediment is dissolved,* and the secretion becomes clear. Put a little of the sediment in a *test tube,* with a solution of potash; upon heat being applied, an odor of ammonia will be perceptible.

General Test for Urates and Uric Acid. Put a little of the sediment on a watch-glass, and dissolve it with a few drops of nitric acid. Evaporate the solution to dryness by heat over a spirit-lamp; when the residue is cold, add a few drops of ammonia, and a purple color will be produced, if uric acid or any of its compounds are present.

Gravel. During what is called a *fit of gravel,* the urine deposits a sediment of one or

* [NOTE.—For a solution of potash, the *liquor-potassæ,* obtained at the druggists, can be used.]

more of the above substances; namely, urea,
uric acid, or urate of ammonia. This difficulty
is attended with pain in the loins, which pain
extends to the testicle. There is frequent
desire to pass water, which operation is attended with more or less pain. There is head-
ache, and a general derangement of the
bowels. The urine is scanty and high-colored.
In some cases there is more or less fever.

Mucus. The urine, left to stand, deposits
a dirty yellow mass, of a tenacious, ropy
character, which *does not mix with the urine
when shaken.* The urine upon being *tested*
does not contain *albumen.*

Earthy Phosphates. In healthy urine, the
amount of earthy phosphates contained is
about one part in one thousand; but in disease
this amount may increase to six parts in one
thousand. The urine, when first voided, is
acid in some cases, but in a short time it becomes alkaline. A short time after it is passed,
a sediment will be deposited. This sediment
in its purity is white and crystalline; but if
the urine contain blood or other coloring
matter, the deposit will be tinged with the
color. If *we heat the urine, the sediment is not
dissolved.* The deposit can be dissolved in
acid, but it is insoluble in water, ammonia, or
in a solution of potash. If the urine be very
acid, the deposit will not take place, as the
earthy phosphates will be held in solution;
but if we add to this acid urine, a large
quantity of ammonia, the deposit will be made.

REMARKS ON THE URINARY DEPOSITS AL-
READY MENTIONED — UREA, URIC ACID,
URATE OF AMMONIA, EARTHY PHOS-
PHATES.

*Sources of Uric Acid, Urea, and Urate of
Ammonia.* The wear and tear of the body, or
the disintegration of the tissues, is one source
of these substances ; therefore, whatever tends
to promote this result will increase the
amount of these elements in the urine. Animal
food introduced into the stomach and badly-
digested, will increase the amount. This may
result if the quantity eaten be too large, or if
not enough exercise be taken to digest it
properly.

Urea, uric acid, and urate of ammonia are
found in excess in the urine in *gout, rheumatism,* diseases of the heart and liver, inflammatory diseases, blows or strain of the loins, and
various diseases of the genital organs—also
when the perspiration is diminished, as in a
common cold.

In nervous affections, as neuralgia and
hysteria, the quantity is diminished in the
urinary secretion.

Sources of Earthy Phosphates. The bladder
is a sac destined to act as a reservoir for the
conservation of urine until such a time as
nature requires its removal from the body.
If it were not for a certain inherent quality in
the bladder itself, which tends to preserve the
secretion, it would undergo decomposition in
that organ before it makes its exit from the

system. This inherent quality depends upon nervous power, which is destroyed if the nerves are paralyzed. Urine, during the process of decomposition, deposits the earthy phosphates. Therefore, when any thing tends to destroy the nervous power of the bladder, the urine is decomposed within that sac, and we have a deposit of earthy phosphates. We may expect a deposit of earthy phosphates in a depressed state of nervous power, from injury of the spine, or if the lining membrane of the bladder be diseased.

Urine containing some one or more substances not found in its healthy state. Sugar is found in healthy urine in VERY small quantities; but in the disease known as Diabetes, it constitutes one of the principal elements. It is the same kind of sugar as that derived from fruit, called grape-sugar. Grape-sugar can be dissolved in strong sulphuric acid, but cane-sugar, if it be put in this acid, is immediately blackened. Grape-sugar is less sweet and less soluble in water than cane-sugar. When sugar is in excess in the urine, the specific gravity is from 1030 to 1050. It has always a pale color. Its odor, when first voided, resembles that of hay. The urine is usually turbid.

Test. Put a little of the urine in a test-tube, and add to it a few drops of a solution of sulphate of copper, at a time, until the urine has a faint blue color; then add a solution of potash, equal in bulk to about one-half the volume of the urine employed. A pale blue precipitate will be thrown down, which will be re-dissolved. Then boil the mixture for a few minutes, and if sugar be present, a reddish-brown precipitate will be thrown down.

Albumen. A large proportion of the fluid portion of the blood consists of albumen. When the blood is accumulated in large amount in any portion or organ of the body, its *serum*, or fluid portion, is effused in the adjoining tissues. A common boil is a good example of this. Thus, when the kidney becomes congested, as in pregnancy, from the pressure of the womb on the veins, or any cause which may lead to congestion of the kidney, we shall find albumen in the urine.

The first stage of Bright's disease of the kidney is a congestion of that gland, and in consequence, albumen is found in the urine. As this disease advances, the structure of the kidney becomes involved, and incapacitated from secreting healthy urine. Albumen is also found in the urine in scarlet fever—of this, hereafter, when treating of Bright's Disease.

Test. Put a little of the urine in a test-tube and add enough nitric acid to render it strongly acid ; then, if the mixture be boiled, and if albumen be present, it will be thrown down in a form resembling the white of an egg boiled. This coagulated albumen (the substance thrown down) will not re-dissolve on heating the urine. It is soluble in a solution of potash.

Blood. If, after the urine has been voided, it be allowed to rest, it will become of a gelatinous consistency; if fibrine, or the coagulable part of the blood be present. If we apply the above test for albumen to the urine, and obtain the deposit as described, it will be found to be colored red if any of the *red particles* of the blood are present.

Test. Add to the urine a strong solution of common salt, and if blood be present, it will become of a light red color.

Bile. The presence of *bile* in the urine gives to that secretion a yellowish brown color, or to any sediment that may be deposited.

Test. Put a little of the urine on a clean, white plate so as to form a layer, then add a few drops of nitric acid, drop by drop. The urine will, if bile be present, present the following colors successively : pale green, violet, pink, yellow — or, in place of these color, showing themselves distinctly—a greenish tint is generally perceptible.

Pus. In urine containing *pus*, we have, after allowing it to stand, a pale green or yellow layer at the bottom. If the urine be shaken, this layer breaks up and *diffuses itself throughout the fluid*, and then again settles. If we find, upon trial, that there is no albumen present in the urine, we can be almost sure that there is no pus, and in cases where there is pus, albumen is also present. The deposit is not dissolved by acetic acid, and by the addition of potash it is rendered more consistent.

Examination of Urinary Deposits. Warm the urine ; if the sediment be dissolved, it is urate of ammonia. If not dissolved, add acetic acid ; if soluble, it is earthy phosphates. If the sediment be not dissolved by the acetic acid, take a new specimen, and test with nitric acid and ammonia ; if a purple color be obtained, it is uric acid.

Examination of the clear portion of the Urine. If, upon the addition of nitric acid, crystals are found, it is owing to the presence of urea. Boil the fluid, and if a precipitate be formed, which is soluble on the addition of nitric acid, it is the earthy phosphates. If the precipitate be not dissolved by nitric acid, after boiling, it is albumen.

Test also, as before explained for mucus, sugar, blood, bile, pus. The mucus and pus need not be tested for in clear urine without deposit.

General indications in the treatment of diseases of the Urinary Organs. We have seen that there is a balance of secretion by which a cersain amount of matter is removed from the tystem. If, from disease or otherwise, one of these functions be prevented from performing its duty, the deficiency is provided for by a compensating increase of secretion from some other outlet of the body. If the perspiration be diminished, the kidney removes the excess of water ; thus supplying the place of the skin. If the function of the liver be deranged, the urine may be colored with bile. For this reason, when we find an apparent deviation from the ordinary quantity, appearance, or composition of the urine, it is not in all cases evidence of disease of the urinary organs, but a simple and healthy action of the kidney in its endeavors to remove the effete matter from the system, or to compensate for other organs which are diseased, and thus preserve the balance of the secretions.

Therefore in the treatment of urinary dis-
orders, our first object should be to look to
the other secretions. We should ascertain
that the *skin* is performing its healthy functions, and if not, apply those means which
will promote its healthy action. This is done
by means of warm baths, friction with a coarse
towel, and exercise—at the same time keep-
ing the body warm by sufficient clothing.
We should also regulate any derangement of
the *digestive organs*—if costive, give a purgative. The *diet* should be regulated. If *uric
acid*, or any of its compounds are present in
large excess, the diet should be mostly vege-
table or fruit. As a drink, a solution of bicar-
bonate of potassa can be taken, one scruple
in a glass of water, with the addition of lemon-
juice. A glass of this can be taken three or
four times in the course of the day. All spirituous liquors, should be avoided.

If the *earthy phosphates* are in excess, attention should be given to restore the general
health by *change of location*, exercise in the
open air, and a plain, nourishing diet, and
thick flannel next the skin. In some cases
tonics would be useful, as some preparations
of iron, or the following : Take of strong nitric
acid, one teaspoonful ; of hydrochloric acid,
two teaspoonfuls ; and of water, fifteen teaspoonfuls—mix these, and take of the mixture
ten drops in a glass of water, three times a day
—to be taken one hour before eating.

Bloody Urine. The presence of blood in the
urine may depend upon a diseased condition
either of the kidney or the bladder, or the in-

jury of those organs from external violence, as a blow on the loins. It may be the result of some gene.al fever or disease, as scarlet fever or scurvy. Also from the irritation of stone.

In case the blood comes from the kidney, there will be pain in the back and loins, while that coming from the bladder will not be attended by these symptoms.

If the blood come from the kidney, the pain in the back and loins not being severe, and no great amount of general derangement, a purge can be given of Epsom salts, and every hour five (5) grains of gallic acid, taken in a little mucilage of gum arabic.

If the patient be weak and debilitated, bladders or bags of pounded ice should be put on the hips and loins and ten (10) drops of muriated tincture of iron should be given every half hour in a little water. As the symptoms improve, less blood being passed, the ice can be left off, and in place of the tincture of iron, ten (10) drops of dilute sulphuric acid in a wine-glass of water, can be given every six hours, with perfect rest.

If there be no pain in the back and loins, and if in voiding the urine some portion of clear blood should pass, the bleeding is from the bladder. A catheter should be introduced into the bladder, and retained, through which cold injections should be made of a solution of one scruple of alum dissolved in one pint of water.

Bleeding from the kidney is rare. It is very difficult to say with *certainty* that it comes from that gland in any given case. The treatment will of course depend upon the severity of the symptoms, and upon the nature of any disease that may have existed previous to the attack. The introduction of a silver catheter is therefore advisable. A gum-elastic bougie in place of metal is better if no surgeon be present.

Stone in the Kidney. There is more or less pain in one or both loins, in case of stone or gravel in the kidney; this, however, is not always constant, as the deposit may exist in that gland for a long period without any particular inconvenience to the patient. Blood is found in the urine, and there is a drawing up or retraction of the testicle. The urine deposits a red sediment. We find more or less general fever, with nausea and vomiting. If the patient take violent exercise, the pain will be increased, while on the other hand, rest will afford relief.

The general health should be attended to. If the pain be severe, and there should be dry skin, heat, and some fever, a few leeches should be put over the seat of pain in the loins; a warm bath should be taken, and some mild cathartic medicine, as magnesia or oil. Injections of warm water will afford relief. For a drink, linseed tea in any quantity desired may be given. The patient can take gentle exercise.

As the stone leaves the kidney in its *passage through the ureter*, the patient will be attacked by a sudden and most severe pain in the groins, which will extend itself to the inside of the thigh and testicle. At the same time there will be faintness, nausea, and vomiting. This may last a few days. It may come on without previous symptoms of derangement.

The patient should take a warm bath, a clyster containing a few drops of laudanum, exercise, large draughts of linseed tea ; and if the symptoms are not relieved, some active cathartic. Occasionally in case of gout, we have the above symptoms, but not the faintness and vomiting. *Treatment :* A purgative, combined with the treatment directed for gout, as wine of colchicum, in doses of ten (10) drops, etc.

Stone in the Bladder. Stone in the bladder arises, as that in the kidney, from a diseased state of the urine, some one or more of its ingredients being in excess. Stone in the bladder may also be caused by some foreign substance introduced into the bladder, and allowed to remain, as a broken end of a catheter or bougie, a piece of a bone, a bodkin, or any other substance introduced into the urethra for surgical purposes, or for the gratification of a morbid taste acquired by the patient. We have removed five inches of fishing line from the bladder by Hurteloups tithotrite ; having undoubtedly been introduced for this purpose. These foreign substances act as a nucleus for the formation of stone, some one or more of the ingredients of the urine accumulating upon their surface.

There is a frequent desire to make water ; and during the act, the urine is of a sudden stopped, which stoppage is relieved by the patient placing himself on his hands and knees.

After having finished voiding the urine, there is more or less pain at the neck of the bladder; pain also at the extreme end of the penis.

Attention should be given to correct the diseased condition of the urine. To prevent pain, a clyster containing a few drops of laudanum. The stone will require to be removed by a proper operation.

Suppression of the Urine. This may result from excess of spirituous liquors taken, or from the kidneys losing their function of secreting during the course of some disease. The patient complains of an uneasy pain in the head and loins; he then becomes drowsy, and finally insensible. He dies in two or three days. There is no treatment of avail.

Retention of Urine; Spasmodic Stricture. The patient tries to urinate, but finds it impossible. This inability to pass the water may continue for several days, the bladder in the mean time becomes distended from the accumulation of urine, and gives rise to general disturbance of the system; a hot skin, an anxious expression of the countenance, and quick pulse. During this period of distension there may be passed from time to time a few drops of urine, but not enough to fully relieve the bladder. As a termination, if relief can not be afforded the patient, the bladder bursts. There may be spasm of the urethra, or swelling and inflammation, accompanied with pain.

The cause of this difficulty may depend upon a diseased condition of the urine, or upon any cause that may render that secretion

irritating, as excessive indulgence in spirituous liquors. It can also result from the ad-
ministration of cantharides either internally
or externally, as in form of blister. Also from
the use of injections in gonorrhœa, provided
they are not suitable to the case, or are used
too often. Cold and wet will sometimes produce this trouble. It is often the consequence
of onanism. This form of retention of urine
is frequently consequent on piles and fissure
of the rectum, to which latter disease women
are very subject.

If the bowels are constipated, a purgative
may be given. A warm hip-bath should be
taken; from ten (10) to twenty-five (25) drops
of laudanum can be given; or tincture of
chloride of iron ten (10) drops every ten minutes. Chloroform or ether inhaled to the extent of anesthesia will often succeed.

If relief is not obtained from the above
treatment, a gum-elastic bougie should be introduced into the bladder to draw off the
water. This operation in some cases will
prove difficult to perform, and in others abso-
lutely impossible. In this case the only means
of relief will be from a surgical operation.*

Occasionally it occurs that women who suf-
fer from nervous disease, find an inability to
pass their water. In this complaint there are
none of the above *severe* symptoms, as the disease is simply nervous, and there is no obstruction to the passage of the urine. Further-
more, from the construction of the parts in

* See Lectures to his private Surgical Class, on stricture, and its cure by the urethrotome, by the author.

women, the bladder will relieve itself, when over-distended. As treatment, a purgative can be given, or a clyster of spirits of turpentine, a spoonful, the yolk of an egg, and one-half pint of mucilage of gum arabic (reduced by the addition of a little water) mixed.

Irritable Bladder. There is occasionally, from a diseased condition of the urine, a frequent desire to pass water, without any violent symptoms. For treatment, avoid all causes that would tend to produce a diseased and irritating state of the urine, as the use of spirituous liquors; attend to the general health, diet, and exercise.

Nocturnal Incontinence of Urine. This habit can be broken up by attention to the general health, diet, and habits, by being awoke at a regular hour in the night to make water. In children it is often relieved by circumcision.

Inflammation of Kidney. Inflammation of the kidney occurs more frequently in the male. When it is found in the female, it is more difficult to determine, as the symptoms of some of the diseases of the womb resemble, to a certain extent, those of inflammation of the kidney. It can also be mistaken for disease of the bowels.

It can be caused by prolonged constipation, from gout, or as a result of disease existing in some of the urinary organs; by external injury, as a blow on the loins, or from overstraining the muscles of the back.

Severe pain is felt in the loins, groins, and

testicle of the side affected, with retraction of the latter gland, (testicle). There is also a numbness of the inside of the thigh. From coughing, sneezing, or any violent motion, the pain is increased: and the same result is obtained from continued pressure. The patient finds temporary relief from rest upon the affected side or back, with legs drawn up. There is general fever, with more or less pain in the abdomen, accompanied with wind.

The disease is distinguished from those of the female generative organs, from the fact that in inflammation of the kidney there is a shooting pain in the direction of the bladder, and a numbness of the thigh ; also by the diseased state of the urine. In males, it can be distinguished from lumbago by the urine, numbness of thigh, and retraction of testicle. The urine passed during the first stage of the disease, will be found to contain albumen and blood. It afterwards becomes pale in color, these substances not being present. The symptoms of improvement are a large discharge of high-colored urine, containing mucus or pus—the pain somewhat relieved, and the skin moist. On the other hand, a sudden cessation of pain, combined with delirium and cold extremities, is an unfavorable symptom.

Leeches should be put on the region of pain, a dose of oil, or some unirritating cathartic given, and for a drink, linseed tea with a little nitre dissolved in it. Perfect rest and quiet. To relieve the pain, a warm hip-bath ; a bag of hops, or poppy-leaves, wrung out in hot water, and placed over the seat of pain ; clysters containing a teaspoonful of laudanum.

Blisters must not be used, as cantharides, when absorbed, they are excessively irritating, often producing inflammation of the bladder. *Bright's or Granular Disease of the Kidney.* The subjects of this disease are generally adults and old people, although it may occur at all periods of life, and in the male and female alike. As a cause, we have a broken-down constitution, which deterioration of health may be the result of debauchery, or intemperance of any kind, impure air, syphilis, or excessive use of mercury. It often follows a chronic inflammation of the kidney, from cold, an injury received, or long continued self-abuse.

This disease may be acute or chronic. The first symptoms of the acute attack may be chill and fever. There will be frequent desire to pass water ; the urine passed will be small in quantity, containing blood and albumen ; a dull pain in the loins, sometimes extending to the testicle ; nausea, and vomiting, general dropsy of the body. The chronic form may come on gradually ; the patient will be obliged to get up in the night to make water; there is a peculiar waxy, pale look in the face, a loss of strength with slight emaciation ; a dry skin, nausea, and great thirst; bad digestion. The disease in a few months may be followed by inflammation, or dropsical diseases of the heart, chest, or lungs, by inflammation of the organs contained in the abdomen, intestines, etc., diarrhœa, or rheumatism. The disease may continue a long period, for months, or even years.

Dr. Grainger Stewart says: " In a large

proportion of cases the symptoms first referred to increase in intensity, the abdominal dropsy becomes so severe as to prevent the free play of the lungs by the descent of the diaphragm, and death from suffocation ensues, or the blood becomes poisoned by the urea, which the diseased kidneys are unable to eliminate; a series of nervous symptoms, varying in their character, but at present grouped under the name uræmic, results, and the patient dies."

Such are the most favorable and the most unfavorable terminations of inflammation of the kidneys; but it must be borne in mind, that in a large number of cases, where the first stage of the disease exists, death occurs not merely from the real malady, but from the combined influence of it and previously existing disease. In such cases death sometimes occurs before dropsy is developed. In others, the dropsy is distinct, but not very prominent. In all, there is a diminution of urine, and more or less copious albumen, whilst tube-casts may be found in almost all the cases.

In a large proportion of cases, instead of complete recovery or death, the symptoms assume a more chronic character; as the quantity of urine rises, its general quality improves, it becomes paler, less albuminous, and entirely free from blood; it still deposits tube-casts, but their quantity is diminished. Yet. the dropsy does not diminish, or alternately increases and is diminished. After a while it remains stationary, the patient becomes drowsy and listless and sinks into unconsciousness—death soon follows.

Happily, new cases which have advanced
to this stage, frequently present more favorable terminations. The urine having increased greatly in quantity, the dropsy gradually
disappears, the amount of albumen in the
urine diminishes, that of urea increases, and
the patient is enabled to return to his duties,
presenting no unfavorable symptom except
albuminuria; and even that may disappear
and the patient be entirely restored.

But, in other cases the recovery is very im-
perfect. The urine is pale, of good quantity,
containing a moderate amount of albumen,
and throws down a large deposit of tube-casts,
but the dropsy never wholly disappears; the
patient cannot return to his duties; if he
does, he relapses, and sooner or later his
symptoms become aggravated and he dies
with increase of dropsy, blood and brain poisoned with urea, and, consequently, uncon-
scious for some time before death, but always
with diminution of urine. In all such cases,
the kidneys are found shrunken or atrophied.

To the correctness of this admirable description we can give our own assurance
from a great number of cases. We have watch-
ed at the bed-side of our friends and patients,
day after day and week after week, and wit-
nessed but little alleviation of the disease
when once fairly established. Our first con-
viction, when we saw the first specimen at the
New York Hospital, in 1830, was correct.
When once disorganization of the kidneys has
fairly commenced, there can be no remedy.
All we can do is to prolong life by the admin-
istration of such tonics as to demand a supply

of food and air, commensurate with the loss of albumen in the urine. After thirty-five years observation of this exhausting and fatal disease, we have found no tonic equal to phosphoric acid and strychnine; it has, undoubtedly, preserved life for years, when the patient would have succumbed under any other tonic; and the reason is obvious; it not only supplies an immediate contractile stimulant to the muscular tissues, but the great element of which the nerves themselves are formed. See article on the blood and nerve-starved condition in " abnormal condition, etc."—Contents at the end of this treatise.

The treatment of this disease is in the first place, to improve the general health by exercise in the open air; a proper nutritive, unstimulating diet; avoiding all fat and oily substances and all pastry and cake. Warm baths should be taken, and sufficient warm under-clothing worn. The bowels should be regulated by cracked wheat and fruits. All spirituous liquors should be avoided. If there be great loss of strength, twenty drops of the U. S. Dispensatory preparation of dilute phosphoric acid, and in extreme debility the sixty-fourth part of a grain of strychnine may be taken three times per day; or citrate of iron and quinine in five (5) grain doses, three times per day. If dropsy be present, active purgatives can be given, with a hot-air bath every three or four days.

When we consider the extensive surface of the skin, and that the amount of perspiration

is quite equal to that passed in the form of
urine from the kidneys, it must be evident
that cold applied to the feet and surface of
the body, in our variable climate, ought to be a
very frequent cause of inflammation of the
kidneys. Cold shuts up the skin, and pro-
duces universal dryness; throwing all the
duty of excreting the surplus fluids of the
body on the kidneys. Every one has felt the
great relief obtained by the bladder from a
free perspiration. Many an attack of inflammation of the kidneys and consequent albuminuria has been averted in a delicate person
—predisposed to it—by a timely resort to a
few warm bricks, and a bowl or two of tea
taken in bed; a hot-air bath, by conducting
the fumes of burning alcohol under the blankets, is very effective. For this purpose a
joint of stove-pipe and a tin cup are all that
is requisite.

Albuminuria often follows diseases of the
skin.

Scarlet fever and erysipelas, diseases in
which the functions of the skin are often
completely interrupted, are both enumerated
by writers, and proved by the experience of
most extensive practitioners, to be fruitful
causes of albuminuria. When after either of
these diseases, a person feels pain in the loins,
has frequent desire to pass water but evac-
cuates little at a time, and that of a high or
smoky color, containing much albumen, and
throwing down many minute long bodies like
worms (casts of albumen from the minute
tubes of the kidneys), the general daily
amount of urine being much diminished and

the feet swelling; he has serious congestion of blood in the kidneys; the vessels are clogged with blood, and albumen is thrown off with the scanty discharge of urine. When the action of the skin is restored by timely remedies, the symptoms subside in a few days, and in a week or two, the urine increases, its color becomes paler, the albumen decreases, the little worm-like tube casts diminish, and the dropsy passes away from the legs and feet, and the person gradually recovers. Such a result is far more frequent in those not reduced from any constitutional cause, but where the eruption has disappeared from cold, suddenly applied; the case then is simply inflammatory congestion of the kidneys, accompanied with its frequent result, effusion of albumen in the urine, and dropsy of the legs. Those cases in which the disease is months in coming on are usually far more serious and fatal; indeed they are often beyond all aid when first submitted to the notice of the physician.

Immoderate flow of Urine. Diabetes. — Is caused from intemperance in the use of spirituous or other liquors; or may be a result of some disease of the urinary organs; or from nervous excitement there may be a much larger quantity of urine passed than natural, attended with loss of health, strength, and flesh, an unpleasant sensation at the stomach, and derangement of the bowels; thirst, with dry skin.

For treatment, the causes should be avoided; less liquor, or fluid, should be drank, attention to the skin, and diet, which should be regulated according to the condition of the urine. In this form there is no *sugar* to be found in the urine.

Another form is that in which, combined with an immoderate flow, *sugar is found in the urine.* This is the more grave disease. As regards the causes of the sugar formation, much has been said, and little decided upon. The disease is the result of an hereditary predisposition, intemperance, mental excitement, or of drinking cold water, the body being over-heated. It is worse in autumn, and in cold and moist places. Consumption is a frequent complication. This form of diabetes is of rare occurrence with children.

This disease, often preceded by some eruption of the skin, commences with frequent desire to pass water, the quantity being increased, great thirst, inordinate appetite, constipation, loss of strength and flesh ; gums are tender, throat dry, the mind is affected, and the breath has the odor of hay. These symptoms increase as the disease advances, when even gallons of urine may be voided during twenty-four hours.

As treatment, all measures should be taken which would improve the general health; such as removal to another location, exercise in the open air, attention to the skin, bowels regulated by aperients, if necessary. Liquids should be drank in moderate quantities only. The diet should be *entirely animal food ;* for a drink, warm beef-tea. The complications can be treated as indications should require ; thus, constipation, by aperients; for pain in the chest or abdomen, a mustard plaster can be applied; if there be great oss of strength, phosphoric acid and strychnine, in appropriate doses three times per day. Dropsy should be treated with purgatives, and attention to the skin.

Inflammation of the Bladder. Inflammation of the bladder may result from a blow inflicted on the lower part of the abdomen; from the irritation of a foreign body, as stone within this sac; or from inflammation of the urethra, (gonorrhœa,) extending to the bladder.

There is frequent desire to make water, the quantity passed being small, and the operation attended with pain. The urine will contain a large quantity of mucus. There will be pain in the lower part of the abdomen, increased by pressure.

When the symptoms are severe, leeches can be applied over the seat of pain, (lower part of pain,) followed by bags of hops, or poppy-leaves, wrung out of warm water, and placed on the part. A warm bath, combined with a clyster of warm water containing thirty drops of laudanum, will be found beneficial. If the pain be not so severe, and mucus be found in large quantities in the urine, take of the root of pareira one ounce, water one pint and one half, boiled to one pint; this can be used as a drink, to advantage; or in place of the above, which however, is the best, copaiba can be taken. The best and most agreeable way of taking copaiba is given in the following formula, which can be prepared by the apothecary:

Take of—Oil of copaiba, seven fluid drachms ;
   Oil of cinnamon, fifteen drops;
   Liquor potassæ, one-half ounce ;
   Magnesia a sufficient quantity;
Triturate, and add—
   Distilled water four ounces ;
Filter, and add—
   Sweet spirits of nitre one-half ounce.
Dose—Teaspoonful three times a day.

If the above measures do not afford relief, an elastic bougie can be introduced into the bladder, and through it injections made into that organ, of a decoction of marsh-mallow, with the addition of a few drops of laudanum.

In chronic cases we have often effected cures by a persistent use of injections, of various strength, of nitrate of silver, given after a thorough washing out of the bladder by a double catheter. As this can only be done by a surgeon, further directions would be superfluous.

Incontinence and Dribbling of Urine. Paralysis of the Bladder. The patient finds it impossible to pass the urine, or else, independent of his control, it dribbles away. This state of things may be brought about by a shock to the nervous system, as a blow on the head or spine ; also, if the bladder be allowed to become over-distended by neglect to void the urine.

In this disease, as well as in every form of sexual debility, we use no other remedy than phosphoric acid and strychnine ; the preparation must be of undoubted accuracy, and should only be given by the educated practitioner. See " Caution " in lectures on irritable bladder and impotence in " Some of the Abnormal conditions which impair virility."

———

SURGICAL HOSPITAL, under the care of EDWARD H. DIXON, M. D., Editor of THE SCALPEL, for the especial treatment of diseases of the Pelvic Viscera. Most cases of Hernia, Hemorrhoids, Fistula, Varicocele, and Stricture now admit of a radical cure in a few days, in place of the palliative treatment of former years.

By Dr. D.'s urethrotome (see " Abnormal Conditions of the Sexual Organs ") Mr. Symes's method of dilating stricture has been reduced to such simplicity that a full-sized catheter may, in most cases, be used at the first attempt. This method is often applicable to Enlarged Prostate. Hemorrhoids and Fistula are often curable without the knife or ligature. Oblique Hernia or Rupture is radically cured by injection, so that the truss may be laid aside. Indeed, the surgery of the Pelvic Viscera may be said to have been entirely changed by the discoveries of the past thirty years.

Over three thousand cases of these diseases have been discharged, cured, within that period by Dr. D.

In all cases, after a personal examination, the case is either pronounced curable or dismissed. Consultations from 8 to 9 A. M., 1 to 3 P. M., and 7 to 9 evenings, at Dr. D.'s residence, No. 42 Fifth Avenue. At all other hours he is at the Hospital. Dr. D. examines and tests microscopically and chemically specimens of urine, not less than four ounces. Specimens may be sent in a phial by express, if inclosed in a tin or wood case.

CONTENTS.

www.ingramcontent.com/pod-product-compliance
Lightning Source LLC
Chambersburg PA
CBHW022029190326
41519CB00010B/1637